Number Systems

Popular Lectures in Mathematics

Survey of Recent East European Mathematical Literature

A project conducted by
Izaak Wirszup,
Department of Mathematics,
the University of Chicago,
under a grant from the
National Science Foundation

Number Systems

84517

Translated and
adapted from the
second Russian
edition by
Joan W. Teller and
Thomas P. Branson

The
University of Chicago
Press
Chicago and
London

The University of Chicago Press, Chicago 60637
The University of Chicago Press, Ltd., London

International Standard Book Number: 0–226–25669–3
Library of Congress Catalog Card Number: 73–89787

S. V. FOMIN, an internationally known functional analyst, is a professor at
Moscow University.

Contents

Preface

The language of numbers, like any language, has its own alphabet. In the language of numbers that is now used virtually worldwide, the alphabet consists of ten digits, 0 through 9. This language is the decimal number system. But this language has not always been used universally. From a purely mathematical point of view, the decimal system has no inherent advantages over other possible number systems; its popularity is due not to mathematical principle, but to a set of historical and biological factors.

In recent times the decimal system has received serious competition from the binary and ternary systems, which are "preferred" by modern computers.

In this pamphlet we will discuss the origin, properties, and applications of various number systems. The reader need not be familiar with mathematics beyond that covered in the high school curriculum.

Two sections (9 and 11) have been added to the second edition, and several minor corrections have been made.

1. Round and Unrounded Numbers

"A man about 49 years old went out for a stroll, walked down the street for about 196 meters, and went into a store; he bought 2 dozen eggs there and then continued walking. . . ." Doesn't that sound a little strange? When we measure something approximately, such as distance or someone's age, we always use round numbers and ordinarily say "200 meters," "50-year-old man," and the like. It is simpler to operate with round numbers: They are easier to remember, and arithmetical computations are easier to perform on them. For example, no one has trouble multiplying 100 by 200 in his head, but multiplying two unrounded three-digit numbers, such as 147 and 343, is so difficult that almost no one can do it without pencil and paper.

In speaking of round numbers, we do not normally realize that the division of numbers into "round" and "unrounded" is dependent on the way in which we are writing the number or, as we usually say, which number system we are using. In investigating this matter, let us first examine the decimal number system, which we all use. In this system every positive integer (whole number) is represented in the form of a sum of ones, tens, hundreds, and so on; that is, in the form of a sum of powers of 10 with coefficients which can assume the values 0 through 9. For example, the notation 2548 denotes the number consisting of 8 ones, 4 tens, 5 hundreds, and 2 thousands; so that 2548 is an abbreviation of the expression

$$2 \cdot 10^3 + 5 \cdot 10^2 + 4 \cdot 10^1 + 8 \cdot 10^0.$$

However, we could, with the same success, represent every number in the form of a combination of powers, not of the number 10, but of any other positive integer (except 1); for example, the number 7. In this

1

system, the so-called heptary number system or the number system with base seven, we would make calculations from 0 to 6 in the usual manner, but we would take the number 7 as a unit of the next order. In our new heptary number system this number is naturally designated by

$$10$$

(a unit of the second order). So that we do not confuse this designation with the decimal number 10, we attach the subscript 7, so that for seven we write

$$(10)_7 \,.$$

Units in the succeeding places serve to denote the numbers 7^2, 7^3, and so forth. They are designated

$$(100)_7 \,, (1000)_7 \,, \text{etc.}$$

We can represent any positive integer by combinations of powers of seven; that is, any positive integer can be expressed in the form

$$a_k \cdot 7^k + a_{k-1} \cdot 7^{k-1} + \cdots + a_1 \cdot 7 + a_0 \,,$$

where each of the coefficients $a_0, a_1, \ldots, a_k$ can assume any whole value from 0 to 6. Just as in the decimal system, it is natural to drop the writing of the powers of the base and to write the number in the form

$$(a_k a_{k-1} \cdots a_1 a_0)_7 \,,$$

again using the subscript to indicate the base of the number system that we are using—in this case, 7.

Let us look again at our example. The decimal number 2548 can be represented as

$$1 \cdot 7^4 + 0 \cdot 7^3 + 3 \cdot 7^2 + 0 \cdot 7 + 0 \,,$$

or, using our notation, as

$$(10300)_7 \,.$$

Thus,

$$(2548)_{10} = (10300)_7 \,.$$

We note that round numbers in the new heptary number system will be completely different from round numbers in the decimal system. For example,

$$(147)_{10} = (300)_7 \,,$$
$$(343)_{10} = (1000)_7$$

(since $147 = 3 \cdot 7^2$ and $343 = 7^3$); at the same time,

$$(100)_{10} = (202)_7 \,,$$
$$(500)_{10} = (1313)_7 \,,$$

and so forth. Therefore, in base seven, multiplying $(147)_{10}$ and $(343)_{10}$ in your head is simpler than multiplying $(100)_{10}$ by $(200)_{10}$. If we used the base seven system, an age of 49 years (and not 50) would undoubtedly be taken as "rounded data," and if we said "98 meters" or "196 meters," it would naturally be taken as an estimate by sight (since $(98)_{10} = (200)_7$ and $(196)_{10} = (400)_7$ are round numbers in base seven). We would count objects by sevens rather than by tens; and so on. In short, if the base seven system were generally accepted, the sentence with which we began would surprise no one.

However, the base seven system is not widely used and can in no way compete with the decimal system, which is used everywhere. What is the reason for this?

2. The Origin of the Decimal Number System

Why does the number 10 play such a privileged role? Someone far removed from these questions would probably answer without thinking, "It is easy to work with the number 10—it is a round number; it is easy to multiply by any number; and, therefore, it is easy to count by tens, hundreds, and so on." But we have already seen that the situation is actually reversed: The number 10 is round only because it is used as the base for the number system. In the transition to another number system, say the heptary system (where ten is written $(13)_7$), the "roundness" of 10 disappears.

The reasons why the decimal system has had such general acceptance are not at all mathematical. The ten fingers of two hands were man's first mechanism for counting. It is convenient to count from one to ten on the fingers. After counting to ten, completely using up our natural "counting apparatus," it is natural to take the number 10 as a new, larger unit (a unit of the next higher order). Ten tens comprise a **unit** of the third order, and so on. Thus, it is because of man's ten-fingered counting that the decimal system, which now seems self-explanatory, originated.

3. Other Number Systems and Their Origins

The decimal system did not always occupy the dominant position. In various historical periods many peoples used number systems other than the decimal. At one time, for example, use of the duodecimal system was rather widespread. Its origin is also undoubtedly connected with counting on the fingers. The four fingers of the hand (excluding the thumb) have a total of 12 phalanges (fig. 1), so that by using the thumb to count off these phalanges in turn, a person could count from 1 to 12. Then 12 is taken as a unit of the next order, and so forth. The duodecimal system has survived in language to this day: Instead of saying "twelve" we often say "a dozen." Many objects (knives, forks, plates, handkerchiefs, and the like) are more often counted by dozens than by tens. (Recall, for example, that a service is, as a rule, for 6 or 12, and much less often for 5 or 10.) Even now the word "gross" is occasionally used, meaning "a dozen dozens" (that is, a unit of the third order in the duodecimal system), and several decades ago it was widely used, especially in the world of commerce. A dozen gross was called a "mass," but now this meaning of the word "mass" is known to few.[1]

Fig..1

The English have unquestionable remnants of the duodecimal system —in their system of measures (for example, 1 foot = 12 inches) and in their monetary system where 1 shilling used to equal 12 pence.

Let us remark that from a mathematical point of view the duodecimal system would have several advantages over the decimal system in that the number 12 is divisible by 2, 3, 4, and 6, while 10 is divisible only by 2 and 5; in general, a large stock of divisors of the base of the number system assures certain conveniences in the use of that system. We shall return to this question in section 7, when we discuss tests for divisibility.

1. However, it is possibly the source of such common expressions as a "mass of men" (compare it with the expression "a thousand men").

In ancient Babylon, where civilization and mathematics were rather advanced, a highly complex base sixty system was used. Historians differ in their explanations of how such a system arose. One of the hypotheses, though it is not particularly believable, states that there was a mingling of two tribes, one of which used the base six system and the other the decimal system. The base sixty system then arose as a compromise between these two systems.

Another hypothesis is based on the Babylonian calculation of the year. Although Babylonian astronomy was sufficiently advanced so that the length of the year could be calculated exactly, the Babylonians found it convenient to designate a period of twelve thirty-day months as a "year," with an extra month added to every sixth year (except for occasional corrections). A year of 360 days naturally leads to the number sixty, since 360 is six times sixty. It has been suggested, however, that the convenience of the 360-day year is itself a result of the Babylonian use of the sexagesimal system. Although the origin of the sexagesimal system remains obscure, its existence and widespread use in Babylon is well established. This system, like the duodecimal, survives to a certain extent in our time (for example, in the division of the hour into 60 minutes and the minute into 60 seconds, and in the analogous system of measuring angles: 1 degree = 60 minutes, 1 minute = 60 seconds). On the whole, however, the Babylonian system, although it did not require the use of sixty different "digits," is rather cumbersome and less convenient than the decimal system.

According to the evidence of Stanley, the explorer of Africa, in a number of African tribes the base five (quinary) system of counting is widely used. The connection of this system with the structure of the human hand is clear enough.

The Aztecs and the Mayas—peoples who lived for many centuries in wide areas of the American continent and who developed a highly advanced civilization which was almost completely destroyed by the Spanish conquistadors in the sixteenth and seventeenth centuries— used the base twenty system. The same base twenty system was used by the Celts, who lived in Western Europe beginning in 2000 B.C. Some traces of the Celts' base twenty system remain in the French language of today. For example, "eighty" in French is *quatre-vingt*—literally "four twenties." The number 20 also used to occur in the French monetary system: The basic monetary unit—the franc—was divided into 20 sous.

Of the four systems of counting cited above (the duodecimal, quinary, sexagesimal, and base twenty), which, along with the decimal, have played an appreciable role in the development of human civilization, all (except the sexagesimal, whose origin is unclear) are connected in some way with counting on the fingers (or on the fingers and toes); that is, like the decimal system, they undoubtedly have an "anatomical" origin.

As the above examples show (their number could have been enlarged), numerous traces of these systems of counting have been preserved in our time in the languages of many peoples, in monetary systems, and in systems of measure. However, in notation and in calculation, we always use the decimal system.

4. Positional and Nonpositional Systems

All the systems of counting which we discussed above are based on one general principle. Some number p is chosen—the base of the number system—and every number N is represented in the form of combinations of its powers with coefficients selected from 0 to $p - 1$, that is, in the form

$$a_k p^k + a_{k-1} p^{k-1} + \cdots + a_1 p + a_0 .$$

Then the number is written in the shortened form

$$(a_k a_{k-1} \cdots a_1 a_0)_p .$$

In this notation the value of each digit depends on the place that the digit occupies. For example, in the number 222, two occurs three times. But the first digit from the right represents two units, the second from the right—two tens (twenty), and the third—two hundreds. (Here we have the decimal system in mind. If we used another number system, say with base p, these three twos would represent the values of 2, $2p$, and $2p^2$, respectively.) Number systems constructed in this way are called *positional*.

Other number systems exist—*nonpositional* number systems constructed on different principles. A well-known example of such a system is the *Roman numeral* system. In that system there are several different basic symbols: the unit I (one), V (five), X (ten), L (fifty), C (hundred), and so on—and every number is represented as a combination of these symbols. For example, in this system the number 88 is

LXXXVIII.

In this system the meaning of a symbol does not depend on the place in which it stands, except when a letter of smaller value is placed to the left of one with larger value. In that case, the position of the letters is important. For example, the expression IV denotes the number four, while VI denotes the number six. In general, though, a given letter will denote the same value regardless of placement; in the representation of the number 88 above, the symbol X occurs three times, each time denoting the same value—ten units.

We often encounter Roman numerals today—on clock faces, for example; they are not used in mathematical practice, however. Positional systems are more convenient because they allow us to represent large numbers using relatively few symbols. A more important advantage of a positional system is the simplicity of performing arithmetical operations on numbers written in such a system. (Try, for comparison, to multiply two three-digit numbers written in Roman numerals.)

From now on we shall consider only positional systems.

5. Arithmetic Operations in Various Number Systems

For numbers written in the decimal system, we use a "columnwise" method for addition and multiplication, and a "diagonal" method for division. These rules are completely applicable to numbers written in any positional system.

Consider addition. In the decimal system, as well as in any other system, we begin by adding the units, then go to the next place, and so on, until we reach the highest available place. We must remember that every time the sum in a preceding place has a result greater than or equal to the base of the number system in which it is written, we must carry over to the next place. For example,

$$(1) \qquad \begin{array}{r} (23651)_8 \\ + (17043)_8 \\ \hline (42714)_8 \end{array}$$

$$(2) \qquad \begin{array}{r} (\ 423)_6 \\ (1341)_6 \\ + (\ 521)_6 \\ \hline (3125)_6 \end{array}$$

We now turn to multiplication. For the sake of definiteness, we choose a specific system, say the hexary (base six). The basis for multiplication of any two numbers is a multiplication table that determines the product of any two numbers smaller than the base of the system. It is not hard to verify that the multiplication table for base six looks like this:

	0	1	2	3	4	5
0	0	0	0	0	0	0
1	0	1	2	3	4	5
2	0	2	4	10	12	14
3	0	3	10	13	20	23
4	0	4	12	20	24	32
5	0	5	14	23	32	41

Here every square contains the product of the numbers of the row and column on whose intersection the square lies, with all numbers written in base six (we have omitted the subscripts in order to make the table more compact).

Using this table, it is easy to multiply by columns numbers containing any number of places. For example,

$$
\begin{array}{r}
(352)_6 \\
\times\ (245)_6 \\
\hline
(3124)_6 \\
(2332)_6 \\
(1144)_6 \\
\hline
(145244)_6
\end{array}
$$

Dividing "diagonally" is also possible in any number system. Consider a problem like the following:

Divide $(120101)_3$ by $(102)_3$.

The solution is

$$
\begin{array}{ll}
(120101)_3 & |(102)_3 \\
\underline{(102)_3} & \quad (1101)_3 \\
(111)_3 & \\
\underline{(102)_3} & \\
(201)_3 & \\
\underline{(102)_3} & \\
(22)_3 &
\end{array}
$$

(Write the divisor, dividend, quotient, and remainder in the decimal system and check the accuracy of the result.)

Problem 1. The following half-written computation was found on the blackboard:

$$\begin{array}{r} 23\text{--}5\text{--} \\ +\ 1\text{--}642 \\ \hline 42423 \end{array}$$

Find out in what number system the addends and the sum were written.

Answer. Base seven.

Problem 2. When we asked a teacher how many pupils were in his class, he answered, "One hundred children—24 boys and 32 girls." At first his answer astonished us, but then we realized that the teacher was simply using a nondecimal system. What system did he have in mind?

The solution to this problem is not complicated. Let x be the base of the number system we are seeking. Then the teacher's words mean that he has x^2 pupils, of whom $2x + 4$ are boys and $3x + 2$ are girls. Thus,

$$2x + 4 + 3x + 2 = x^2 ,$$

or

$$x^2 - 5x - 6 = 0 ,$$

yielding

$$(x - 6)(x + 1) = 0 ,$$

or, by the quadratic formula,

$$x = \frac{5 \pm \sqrt{(25 + 24)}}{2} = \frac{5 \pm 7}{2} ;$$

either method yields

$$x_1 = 6 , x_2 = -1 .$$

Since -1 cannot be the base of a number system, $x = 6$. Thus, the teacher's answer was in the base six system, and he had 36 pupils—16 boys and 20 girls.

6. Translating Numbers from One System to Another

How do we translate a number written in one system, say the decimal, into another system, say the base seven system? We already know that to write a number A in base seven is to represent it as the sum:

$$A = a_k \cdot 7^k + a_{k-1} \cdot 7^{k-1} + \cdots + a_1 \cdot 7 + a_0 .$$

Consequently, in order to find the representation of the number A to the base seven, we need to find the coefficients $a_0, a_1, \ldots, a_k$, each of which can be some digit between 0 and 6, inclusive. We divide our number A by 7 (in integers). The remainder in this division is clearly equal to a_0, since, in the representation of A, all the terms except the last are evenly divisible by 7. Then let us take the quotient obtained from dividing the number A by 7, and again divide it by 7. This newly obtained remainder will be equal to a_1. We continue this process and find all the digits $a_0, a_1, \ldots, a_k$ in the base seven representation of A, in the form of successive remainders obtained by dividing it repeatedly by 7, as described above. Consider, for example, the number

$$(3287)_{10} .$$

Dividing it by 7, we get a quotient of 469 and a remainder of 4. Consequently, written to the base 7, the number 3287 has its last digit equal to 4. To find the next-to-last digit, we divide our quotient 469 again by 7. We get a quotient of 67 and a remainder of 0. Consequently, the next-to-last digit of the number 3287 to the base 7 is 0. Further, we divide 67 by 7, obtaining 9 with a remainder of 4. This remainder of 4 represents the third digit from the end of 3287 written to the base seven. Finally, we divide the last quotient 9 by 7, getting a remainder of 2 and a quotient of 1. The remainder of 2 gives us the fourth digit from the end in the desired notation, and the quotient of 1 (which we can no longer divide by 7) represents the fifth digit from the end (the first digit). Thus,

$$(3287)_{10} = (12404)_7 .$$

The right side of this equation is an abbreviation of the expression

$$1 \cdot 7^4 + 2 \cdot 7^3 + 4 \cdot 7^2 + 0 \cdot 7 + 4 ,$$

just as $(3287)_{10}$ is an abbreviation of the expression

$$3 \cdot 10^3 + 2 \cdot 10^2 + 8 \cdot 10 + 7 .$$

The computations that we used for translating from the decimal representation of the number 3287 into its representation to the base seven are conveniently arranged as follows:

```
3287 | 7
  4   469 | 7
        0   67 | 7
              4   9 | 7
                    2   1
```

Everything that we have said above clearly applies not only to the base seven system, but to any other such system. A general rule for obtaining the representation of some number A in the number system with base p can be formulated in the following way: Divide the number A by p in integers; the remainder thus obtained will be the last digit of the base p representation of the number A. Dividing the quotient obtained from this division again by p leaves us a second remainder; this will be the digit that occupies the next-to-last place; and so forth. The process continues until we obtain a quotient smaller than p, the base of the system. That quotient is the digit that occupies the highest place.

Let us consider one more example. The problem is to write the number 100 in binary notation. We obtain:

$$
\begin{array}{c c c c c c c}
100 & | & 2 & & & & \\
0 & 50 & | & 2 & & & \\
& 0 & 25 & | & 2 & & \\
& & 1 & 12 & | & 2 & \\
& & & 0 & 6 & | & 2 \\
& & & & 0 & 3 & | & 2 \\
& & & & & 1 & 1
\end{array}
$$

that is,

$$(100)_{10} = (1100100)_2 .$$

One constantly encounters the problem of translating numbers from the decimal to the binary system when working with computers, a subject about which we shall have more to say later.

In the examples we have considered, the original number system has been the decimal system. We can, however, translate numbers from any given system to any other by the same means. To do so, we need only note that the process of successive divisions carried out in the above examples can also be carried out in any base in which we are given the original number representation.

Problem. Let us assume we have a scale (with two pans) and weights of 1 gram, 3 grams, 9 grams, 27 grams, and so on (one object of each weight). Using only this equipment, is it possible to weigh any mass to within an accuracy of one gram? The answer is yes. We shall present the solution here, relying on the representation of positive integers in the ternary system. Suppose the object that we wish to weigh weighs A grams (taking A as an integer). We can write the number A in the ternary system as

$$A = (a_n a_{n-1} \cdots a_1 a_0)_3 \,,$$

that is,

$$A = a_n \cdot 3^n + a_{n-1} \cdot 3^{n-1} + \cdots + a_1 \cdot 3 + a_0 \,,$$

where the coefficients $a_0, a_1, \ldots, a_n$ can assume values of 0, 1, or 2.

It is possible, however, to write each number in the ternary system somewhat differently, so as to use the digits 0, 1, and -1 (instead of 0, 1, and 2). We utilize such a system as follows: We translate the number A from the decimal to the ternary system, using the method of successive divisions that we described earlier, except that every time we divide by 3 and get a remainder of 2, we will increase the quotient by 1, leaving a remainder of -1.

As a result, we obtain the number A in the form of a sum:

$$A = b_m \cdot 3^m + b_{m-1} \cdot 3^{m-1} + \cdots + b_1 \cdot 3 + b_0 \,,$$

where each of the coefficients $b_m, b_{m-1}, \ldots, b_0$ can assume a value of 0, 1, or -1. For example, the number 100, which, in the usual ternary notation, would have been 10201, would have the form $11(-1)01$ in this variation, since

$$(100)_{10} = (10201)_3 = 3^4 + 3^3 - 3^2 + 1 \,.$$

Now, we put the mass weighing A grams on the first pan of the scale, and we put a weight of one gram on the second pan if $b_0 = 1$, and on the first pan if $b_0 = -1$. (If $b_0 = 0$, we do not use the first weight.) Continuing, we put the 3 gram weight on the second pan if $b_1 = 1$, and on the first if $b_1 = -1$, and so on. It is easy to see that if we arrange the weights in this manner we can balance the weight A. And so, with the help of weights of mass 1, 3, 9, and so on, it is possible to balance any integral mass on the scales. If the weight of the mass is unknown, we choose a distribution of weights that balances the mass and thus determines the weight.

Let us clarify our discussion with an example. Suppose we have a mass weighing 200 grams. Translating 200 into ternary notation in the usual way, we obtain

$$
\begin{array}{r|l}
200 & 3 \\ \hline
2 \quad 66 & 3 \\ \hline
0 \quad 22 & 3 \\ \hline
1 \quad 7 & 3 \\ \hline
1 \quad 2 &
\end{array}
$$

Consequently,

$$(200)_{10} = (21102)_3 ,$$

or, in greater detail,

$$200 = 2 \cdot 3^4 + 1 \cdot 3^3 + 1 \cdot 3^2 + 0 \cdot 3 + 2 .$$

If 200 is translated into ternary notation of the second type, using -1 and not 2, we obtain

$$
\begin{array}{r|l}
200 & 3 \\ \hline
-1 \quad 67 & 3 \\ \hline
1 \quad 22 & 3 \\ \hline
1 \quad 7 & 3 \\ \hline
1 \quad 2 & 3 \\ \hline
-1 \quad 1 &
\end{array}
$$

that is,

$$200 = 1 \cdot 3^5 - 1 \cdot 3^4 + 1 \cdot 3^3 + 1 \cdot 3^2 + 1 \cdot 3 - 1 .$$

(The validity of this last equality is easily verified by direct calculation.)

So, in order to balance a mass of 200 grams placed on a pan of the scales, we need to put weights of 1 gram and 81 grams on the same pan and weights of 3, 9, 27, and 243 grams on the other pan.

7. Tests for Divisibility

There are simple tests that permit us to determine whether a given number is divisible, for example, by 3, 5, 9, and so forth. Let us recall those tests:

1. *Test of divisibility by 3.* A number is divisible by 3 if and only if the sum of its digits is divisible by 3. For example, the number 257802 (in which the sum of digits is $2 + 5 + 7 + 8 + 0 + 2 = 24$) is divisible by 3, but the number 125831 (in which the sum of digits is 20) is not divisible by 3.

2. *Test of divisibility by 5.* A number is divisible by 5 if and only if its last digit is either 0 or 5 (that is, if and only if 5 divides the number of units in the last place).

3. *Test of divisibility by 2.* This test is analogous to the immediately preceding one: A number is divisible by 2 if and only if 2 divides the number of units in the last place.

4. *Test of divisibility by 9.* This is analogous to the test of divisibility by 3: A number is divisible by 9 if and only if the sum of its digits is divisible by 9.

Proof of the validity of these tests presents no difficulty. Let us examine, for example, the test of divisibility by 3. It is based on the fact that the units in each place in the decimal system (that is, the numbers 1, 10, 100, 1000, and so on) leave a remainder of 1 when divided by 3. Therefore, since every number

$$(a_n a_{n-1} \cdots a_1 a_0)_{10} \, ,$$

that is, every number

$$a_n \cdot 10^n + a_{n-1} \cdot 10^{n-1} + \cdots + a_1 \cdot 10 + a_0$$

can be written in the form

$$(a_n + a_{n-1} + \cdots + a_1 + a_0) +$$
$$[a_n(10^n - 1) + a_{n-1}(10^{n-1} - 1) + \cdots + a_1(10 - 1) + a_0(1 - 1)],$$

and since $(10^k - 1)$ for $k = 0, \ldots, n$ is divisible by 3, we may write our number in the form

$$(a_n + a_{n-1} + \cdots + a_1 + a_0) + B \, ,$$

where B is evenly divisible by 3. It is clear, then, that the number

$$a_n \cdot 10^n + a_{n-1} \cdot 10^{n-1} + \cdots + a_1 \cdot 10 + a_0$$

is divisible by 3 if and only if 3 divides the number

$$a_n + a_{n-1} + \cdots + a_1 + a_0 \, .$$

For example, the decimal number 4851 can be written

$$4851 = 4 \cdot 1000 + 8 \cdot 100 + 5 \cdot 10 + 1$$
$$= 4 \cdot (999 + 1) + 8 \cdot (99 + 1) + 5 \cdot (9 + 1) + 1$$
$$= 4 + 8 + 5 + 1 + (4 \cdot 999 + 8 \cdot 99 + 5 \cdot 9)$$
$$= 4 + 8 + 5 + 1 + B,$$

where B is divisible by 3. Thus, since 3 divides $4 + 8 + 5 + 1 = 18$, 3 must divide 4851.

The test of divisibility by 5 is based on the fact that the number 10—the base of the number system—is divisible by 5; therefore, all the powers of ten from the first on are divisible by 5. Therefore, if a number is to be divisible by 5, its last digit must yield a remainder of 0 on division by 5. The test for divisibility by 2 has the same basis: A number is even if and only if its last digit is even.

The test of divisibility by 9, like the test of divisibility by 3, is based on the fact that every number of the form 10^k leaves a remainder of 1 when divided by 9.

From the discussion, it is clear that all these tests are based on decimal representation of integers, and that they are, generally speaking, inapplicable if we use a different number system. For example, the number 86 is written to the base 8 in the form

$$(126)_8$$

(since $86 = 8^2 + 2 \cdot 8 + 6$). The sum of the digits is 9, but 86 is divisible by neither 3 nor 9.

However, in any positional system it is possible to formulate tests for divisibility by various numbers. Let us consider a few examples.

We shall write numbers in the duodecimal system and formulate, for that notation, a test for divisibility by 6. Since the number 12—the base of the number system—is divisible by 6, a number written in the duodecimal system is divisible by 6 if and only if 6 divides its last digit. (We have here the same situation as for divisibility by 2 and by 5 in the decimal system.)

Since the numbers 2, 3, and 4 also divide the number 12, the following divisibility tests are valid: A number to the base 12 is divisible by 2, 3, or 4, respectively if and only if its last digit is divisible by 2, 3, or 4, respectively.

We leave it to the reader to check the validity of the following divisibility tests in the duodecimal system:

a. The number $A = (a_n a_{n-1} \cdots a_1 a_0)_{12}$ is divisible by 8 if and only if the number $(a_1 a_0)_{12}$ (formed by the last two digits of A) is divisible by 8. (*Hint:* 8 is a divisor of $12^2 = 144$, so that all the powers of 12 from the second on are divisible by 8.)

b. The number $A = (a_n a_{n-1} \cdots a_1 a_0)_{12}$ is divisible by 9 if and only if the number $(a_1 a_0)_{12}$ (formed by the last two digits of A) is divisible by 9. (*Hint:* 9 divides 144.)

c. The number $A = (a_n a_{n-1} \cdots a_1 a_0)_{12}$ is divisible by 11 if and only if the sum of its digits (the number $a_n + a_{n-1} + \cdots + a_1 + a_0$) is divisible by 11. (*Hint:* $12^k - 1$ is divisible by 11 for $k = 0, \ldots, n$, since $12^k - 1 = 11 \cdot 12^{k-1} + 11 \cdot 12^{k-2} + \cdots + 11 \cdot 12 + 11$.)

Let us consider two more problems connected with the divisibility of numbers.

Problem 1. The number $(3630)_p$ (written in base p) is divisible by 7. What is p, and what is the decimal representation of the number A if we know that $p \leq 12$? Will the problem's solution be unique if the condition $p \leq 12$ is not satisfied?

Solution. Since 7 is a *prime number* (that is, a number whose only positive integral divisors are itself and one), it can be shown that if 7 divides the product ab and 7 does not divide a, then 7 divides b. To apply this information to the problem, we may write

$$(3630) = 3p^3 + 6p^2 + 3p = 3p(p + 1)^2 \,.$$

Since 7 does not divide 3, 7 divides $p(p + 1)^2$. Since 7 is a prime number, divisibility of $(p + 1)^2$ by 7 would imply the divisibility of $p + 1$ by 7. Obviously, 7 cannot divide both p and $p + 1$. Applying this information, we know that 7 must divide either p or $p + 1$. If $p \leq 12$, p must be either 6 or 7. However, in base 6, the digit string 3630 is meaningless; therefore, $p = 7$. From this it is easy to calculate $A = (1344)_{10}$. If the condition $p \leq 12$ is not satisfied, p may be any number of the form $7k$ or $7k - 1$, where $k = 1, 2, \ldots$ (except for $7 \cdot 1 - 1 = 6$).

Problem 2. Prove that the number

$$(a_n a_{n-1} \cdots a_1 a_0)_p \,,$$

that is, the number

$$a_n \cdot p^n + a_{n-1} \cdot p^{n-1} + \cdots + a_1 \cdot p + a_0 \,,$$

is divisible by $p - 1$ if and only if $p - 1$ divides the sum

$$a_n + a_{n-1} + \cdots + a_1 + a_0 \,.$$

(Compare this general problem with the divisibility test for 9 in the decimal system and with the divisibility test for 11 in the duodecimal system.)

8. The Binary System

The smallest integer that can be used as the base of a number system is 2. The binary (base 2) system is one of the very oldest. It is encountered, although in a very incomplete form, among several Australian and Polynesian tribes. The convenience of the system is its extraordinary simplicity. In the binary system we have only two digits, 0 and 1, and the number 2 is a unit of the second order. The rules for operations on binary numbers are very simple. The basic rules for addition are given by

$$0 + 0 = 0, \quad 0 + 1 = 1, \quad 1 + 1 = (10)_2,$$

and the multiplication table for the binary system has the form

	0	1
0	0	0
1	0	1

A slight disadvantage of the binary system arises as a result of the small size of the base. This means that writing even moderately large numbers requires the use of many places. For example, the number 1000 is written in the binary system in the form

$$1111101000,$$

using ten digits. However, this disadvantage is often compensated for by the convenience of the binary system in modern technology, especially in the use of computers.

We shall talk about the technological applications of the binary system later, but, for the present, let us consider two problems connected with binary number notation.

Problem 1. I am thinking of some base 10 integer between 0 and 1000. Can you find out what the number is, asking no more than ten "yes or no" questions? This problem is completely solvable.

One possible series of questions automatically leading to success is:

First question: "Is the number you are thinking of evenly divisible by 2?" If the answer is yes, we write down the number zero; if not, we write down the number one. (In other words, we write down the remainder obtained by dividing the "secret" number by 2.)

Second question: "Divide the quotient, which you obtained from the first division, by 2. Is it evenly divisible?" Again, if the answer is yes, we write a zero, and if it is no, we write a one.

Each succeeding question will be of the same form, that is, "Divide the quotient, which was obtained from the previous division, by 2. Is it evenly divisible?" Each time, we write a zero if the answer is affirmative and a one if the answer is negative.

Using this procedure 10 times, we obtain 10 digits, each of which is either zero or one. It is easy to see that these digits form the binary representation of the desired number in reverse order. Actually, our system of questions reproduces the procedure by which a number is translated into the binary system. The ten questions are enough because every number from 1 to 1000 can be written in binary notation using no more than ten places (since $1024 = 2^{10}$). If the intended number had been written in binary notation in the first place, it would have been clear how our ten questions were functioning: We were actually asking whether each of the digits was a zero or a one.

Let us consider another problem which is closely related to this one.

Problem 2. I have seven tables, each of which contains a chessboard of 64 squares (fig. 2).

In each square is written a number from 1 to 127. Choose one of these numbers, and tell me in which of the tables (they are numbered from 1 to 7) that number is located. I can name the number. How?

Here is the solution to this uncomplicated problem:

Let us write every number from 1 to 127 in binary notation. None of these representations has more than seven places, since $127 = (1111111)_2$. We put a number A in the kth table ($k = 1, 2, \ldots, 7$) if, in its binary representation, the kth place from the right has a 1, and we do not write it there if the kth place is occupied by a zero. For example, the number 57, which is written in binary notation as

$$0111001,$$

would be contained in the first, fourth, fifth, and sixth tables, the number 1 only in the first table, the number 127 in all of the tables, and so on. In this way, if we know in which tables a number is contained, we know its binary representation. All we need do is translate it into the decimal system.

1	3	5	7	9	11	13	15
17	19	21	23	25	27	29	31
33	35	37	39	41	43	45	47
49	51	53	55	57	59	61	63
65	67	69	71	73	75	77	79
81	83	85	87	89	91	93	95
97	99	101	103	105	107	109	111
113	115	117	119	121	123	125	127

1

2	3	6	7	10	11	14	15
18	19	22	23	26	26	30	31
34	35	38	39	42	43	46	47
50	51	54	55	58	59	62	63
66	67	70	71	74	75	78	79
82	83	86	87	90	91	94	95
98	99	102	103	106	107	110	111
114	115	118	119	122	123	126	127

2

4	5	6	7	12	13	14	15
20	21	22	23	28	29	30	31
36	37	38	39	44	45	46	47
52	53	54	55	60	61	62	63
68	69	70	71	72	77	78	79
84	85	86	87	92	93	94	95
100	101	102	103	108	109	110	111
116	117	118	119	124	125	126	127

3

8	9	10	11	12	13	14	15
24	25	26	27	28	29	30	31
40	41	42	43	44	45	46	47
56	57	58	59	60	61	62	63
72	73	74	75	76	77	78	79
88	89	90	91	92	93	94	95
104	105	106	107	108	109	110	111
120	121	122	123	124	125	126	127

4

16	17	18	19	20	21	22	23
24	25	26	27	28	29	30	31
48	49	50	51	52	53	54	55
56	57	58	59	60	61	62	63
80	81	82	83	84	85	86	87
88	89	90	91	92	93	94	95
112	113	114	115	116	117	118	119
120	121	122	123	124	125	126	127

5

32	33	34	35	36	37	38	39
40	41	42	43	44	45	46	47
48	49	50	51	52	53	54	55
56	57	58	59	60	61	62	63
96	97	98	99	100	101	102	103
104	105	106	107	108	109	110	111
112	113	114	115	116	117	118	119
120	121	122	123	124	125	126	127

6

64	65	66	67	68	69	70	71
72	73	74	75	76	77	78	79
80	81	82	83	84	85	86	87
88	89	90	91	92	93	94	95
96	97	98	99	100	101	102	103
104	105	106	107	108	109	110	111
112	113	114	115	116	117	118	119
120	121	122	123	124	125	126	127

7

Fig. 2

The question can be reversed: Choose a number from 1 to 127, and I will tell you in which of the tables in figure 2 it is located and in which it is not. To answer the question, all we need do is translate the given number into the binary system (with a little practice this is not difficult to do in your head) and then simply name those places that are occupied by a 1.[2]

The above discussion leaves one unanswered question, however. Why are there exactly sixty-four numbers in each table? Let us consider table k ($k = 1, 2, \ldots, 7$) in which we have written all numbers from 1 through 127 which have a one in the kth place from the right. To each of these numbers, there corresponds a *unique* number which is derived from the first by substituting 0 for 1 in the kth place. This second number, of course, will *not* be in table k; yet it will be between 0 and 127 inclusive (and will be zero only when the original number has 1 *only* in the kth place). Furthermore, all the numbers not in table k derived by this correspondence will be distinct (as is easy to verify); and each number not in table k can be derived from some number in table k by the correspondence. Thus, if there are n numbers in table k, there must be $n - 1$ numbers not in table k (discounting zero, and allowing only numbers from 1 through 127), so that the following deduction holds:

$$n + (n - 1) = 127 \; ;$$
$$2n - 1 = 127 \; ;$$
$$2n = 128 \; ;$$
$$n = 64 \; .$$

Since this is true for all $k = 1, 2, \ldots, 7$, there must be exactly 64 numbers in each table.

9. The Game of Nim

A game called "Nim" was popular in ancient China. It involves three piles of stones; two players alternate in taking stones from the piles; in each turn a player can take any nonzero number of stones from any pile (but only from one). The winner is the one who takes the last stone.

Nowadays, more convenient objects are used in place of stones—for example, matches. The problem lies in clarifying the optimal strategy for each player.

2. In each of the above-mentioned tables, the numbers are written in order of increasing size, making the structure of these tables fairly easy to discover. However, within each of the seven tables, the numbers can be rearranged quite arbitrarily, hiding the method by which the tables were constructed.

The binary system is helpful in solving this problem. Suppose that there are a, b, and c matches in the three piles. We write the numbers a, b, and c in binary notation:

$$a = a_m \cdot 2^m + a_{m-1} \cdot 2^{m-1} + \cdots + a_1 \cdot 2 + a_0 \,,$$
$$b = b_m \cdot 2^m + b_{m-1} \cdot 2^{m-1} + \cdots + b_1 \cdot 2 + b_0 \,,$$
$$c = c_m \cdot 2^m + c_{m-1} \cdot 2^{m-1} + \cdots + c_1 \cdot 2 + c_0 \,.$$

If necessary, we put zeros in front of the numbers which have fewer digits. In this way, each of the digits $a_0, b_0, c_0, \ldots, a_m, b_m, c_m$ can be equal to either 0 or 1, with at least one of the digits a_m, b_m, and c_m (though not necessarily all) different from zero. The player who goes first may replace one of the numbers a, b, or c with any smaller number. Suppose that he decides to take matches from the first pile, that is, to change the number a. This means that some of the digits $a_0, a_1, \ldots, a_m$ will be changed. Analogously, in taking the matches from the second pile, the player would change at least one of the digits $b_0, \ldots, b_m$; and, taking matches from the third pile, he would change at least one of the digits $c_0, \ldots, c_m$.

Now consider the sums

$$a_m + b_m + c_m \,, a_{m-1} + b_{m-1} + c_{m+1} \,, \ldots, a_0 + b_0 + c_0 \,. \qquad (*)$$

Each of these sums can equal 0, 1, 2, or 3. If at least one of these sums is odd (that is, equals 1 or 3), then the player with the first turn is assured of victory. In fact, let $a_k + b_k + c_k$ be the first (counting from the left) of the sums in $(*)$ which are odd. Then at least one of the three numbers a_k, b_k, and c_k is equal to 1. Assume, without loss of generality, that $a_k = 1$. Then the first player can take from the first pile any number of matches such that the coefficients $a_m, \ldots, a_{k+1}$ do not change, a_k is equal to 0, and every one of the coefficients $a_{k-1}, \ldots, a_0$ can take whatever value (0 or 1) the player desires. Thus, the player can take a number of matches from the first pile such that all the sums

$$a_{k-1} + b_{k-1} + c_{k-1}, \ldots, a_0 + b_0 + c_0$$

become even.

In other words, the first player can arrange it so that after his turn all the sums in $(*)$ have become even. The second player, in making his move, cannot help but change the evenness of at least one of the sums, since he must change at least one digit in some number, but in only one number. This means that after his turn we again have the situation in which at least one of the sums in (∗) is odd. The first player, in his next move, can again even out all the sums. And so, after every turn of the first player, all the sums in (∗) are even, and after every turn of the second player, at least one of these sums is odd. Since the total number of matches decreases after every turn, we eventually reach the situation where all the sums in (∗) are zero—there are no matches left. Since all sums are even when and only when the first player has just taken his turn, the first player must have taken the turn that reduced all the sums to zero, and so he must have taken the last match; he has won.

For example, suppose that initially $a = 7$, $b = 6$, and $c = 2$. We would then write

$$a = (111)_2 ;$$
$$b = (110)_2 ;$$
$$c = (010)_2 .$$

The sums of interest would be

$$a_2 + b_2 + c_2 = 1 + 1 + 0 = 2 ,$$
$$a_1 + b_1 + c_1 = 1 + 1 + 1 = 3 ,$$
$$a_0 + b_0 + c_0 = 1 + 0 + 0 = 1 .$$

The "first" odd sum is $a_1 + b_1 + c_1 = 3$. The first player may then decide to draw from the first pile. In doing so, he should change a_1 from 1 to 0. But he must also arrange for $a_0 + b_0 + c_0$ to be even, so he must also change a_0 from 1 to 0. The result is the subtraction of $(011)_2 = 3$ stones from the first pile, leaving

$$a = 4 = (100)_2 ;$$
$$b = 6 = (110)_2 ;$$
$$c = 2 = (010)_2 .$$

The sums

$$a_2 + b_2 + c_2 = 2 ,$$
$$a_1 + b_1 + c_1 = 2 ,$$
$$a_0 + b_0 + c_0 = 0$$

are then all even.

The second player may decide to draw three stones from the first pile (it doesn't matter what he decides; if the first player knows the optimal strategy, he has no chance). This would leave

$$a = 1 = (001)_2 \; ;$$
$$b = 6 = (110)_2 \; ;$$
$$c = 2 = (010)_2 \; ,$$

and

$$a_2 + b_2 + c_2 = 1 \; ,$$
$$a_1 + b_1 + c_1 = 2 \; ,$$
$$a_0 + b_0 + c_0 = 1 \; .$$

The first player must then decide to draw from pile two, in order to change b_2 from 1 to 0 (the only way to even out $a_2 + b_2 + c_2$ without adding stones). In doing so, he must change b_0 from 0 to 1, while leaving b_1 fixed. In other words, he must change b from $(110)_2 = 6$ to $(011)_2 = 3$ by drawing 3 stones. This leaves

$$a = 1 = (01)_2 \; ;$$
$$b = 3 = (11)_2 \; ;$$
$$c = 2 = (10)_2 \; .$$

Suppose the second player decides to simplify the game by removing the stone from the first pile (again, it doesn't matter what he decides). This leaves (omitting the first pile)

$$b = 3 = (11)_2 \; ;$$
$$c = 2 = (10)_2 \; ,$$

and

$$b_1 + c_1 = 2 \; ,$$
$$b_0 + c_0 = 1 \; .$$

The first player then removes one stone from the second pile, so that

$$b = c = 2 = (10)_2 \; ;$$
$$b_1 + c_1 = 2 \; ;$$
$$b_0 + c_0 = 0 \; .$$

At this point, the second player will not remove one of the piles, for that would mean instant defeat. Instead, he draws one stone, say from the last pile, leaving

$$b = 2 = (10)_2 \; ;$$
$$c = 1 = (01)_2 \; ;$$
$$b_1 + c_1 = 1 \; ;$$
$$b_0 + c_0 = 1 \; .$$

The first player must change both the first and second places in one of the numbers; this can be accomplished only by changing b from $(10)_2 = 2$ to $(01)_2 = 1$. Then

$$b = c = 1 \, ,$$

and the second player must remove one of the stones, after which the first player removes the other and wins.

If all the sums in (∗) are initially even, the first player's first turn makes at least one of the sums odd, allowing the second player to win by using the above strategy.

Thus, if the optimal strategy is known to both players, the numbers a, b, and c completely determine the result.

Of course, three numbers that would give the second player victory rarely occur, and so in the long run the first player will do far better than the second. For example, there are eight ways to divide ten matches $(a + b + c = 10)$ into three piles. Seven of these arrangements determine victory for the first player, while only one favors the second.

At least one important question is raised, however. Could more "optimal" strategies be devised, using number systems to bases other than 2? For example, could the ternary system be used so that the first player's object would be to make the sums of corresponding digits all divisible by 3? The answer is no since, when a base p number system is used, the "optimal" strategy breaks down as soon as the combined total of the number of matches in the three piles becomes less than p. In this situation, it is impossible for the first player to arrange for the sum of the digits in the units' place to be divisible by p (unless two piles have been exhausted). If, in addition, not all of the binary sums of interest were even, the second player could apply the binary strategy to win. Such a situation could occur if the first player, using the ternary strategy, left exactly one match in each pile after his turn. The remaining turns of both players would then be determined, and the second player would draw the last match.

The effectiveness of the binary system in this application lies in the fact that if fewer than two matches remain in all three piles combined, then only one match remains, and the first player can win simply by drawing the match.

10. The Binary Code and Telegraphy

One of the oldest technical uses of the binary system is the telegraph code. We write the letters of the alphabet and a space denoted by "——", numbering them from 0 to 26:

——	A	B	C	D	E	F	G	H	I	J	K	L	M
0	1	2	3	4	5	6	7	8	9	10	11	12	13

N	O	P	Q	R	S	T	U	V	W	X	Y	Z
14	15	16	17	18	19	20	21	22	23	24	25	26

Each of these numbers (0 through 26) can be written in the binary system using no more than five digits, since $2^5 = 32$. We obtain

——	0	0	0	0	0
A	0	0	0	0	1
B	0	0	0	1	0
.	.	.	.	.	.
.	.	.	.	.	.
Z	1	1	0	1	0

Suppose that we have five conductors joining two points. Then each five-digit figure representing a unique letter of the alphabet can be sent from one point to the other using a definite combination of electrical impulses: Say we let no signal stand for 0 and an impulse on the appropriate conductor stand for 1. At the point of reception this combination of impulses will set into operation a printing apparatus which will print the letter corresponding to the given combination of impulses (and thus to the given binary number).

The telegraph is, in principle, a combination of two apparatuses: an initial mechanism which translates the message into a system of impulses to be sent across the connecting lines, and a receiving mechanism which translates the impulses into a combination of letters via a printing mechanism.[3]

3. We have been speaking of two points linked by five conductors. However, we can manage with only one conductor by transmitting each letter as a succession of five binary digits (impulse or no impulse).

In this way, the ease of translating binary numbers into a system of electrical impulses leads to a useful application of the binary system in telegraphy.[4]

11. The Binary System—A Guardian of Secrets

Telegraph and radio-telegraph provide for fast transfer of information. However, telegraph messages are easily intercepted, and sometimes, especially in military matters, information must be made accessible only to the intended recipient of the message. Therefore, we must resort to coding methods.

We have all, at some time, used codes and conducted "secret correspondences." The simplest type of code is constructed by representing each letter of the alphabet by some symbol: another letter, a number, a convenient mark, and so on. Such codes are often important in detective and mystery stories; for example, Conan Doyle's "The Adventure of the Dancing Men" and Jules Verne's *Journey to the Center of the Earth*. Such codes are easily broken.

Any language, including Russian or English, has a definite structure: Some letters and letter combinations occur often, some less often, and others (for example, a *w* following a *q* in English) not at all. This structure is independent of the choice of alphabetic symbols, and so it remains after coding, allowing us to discover the coding system and the actual message. Even coding systems far more complex than those of this type yield their secrets to an experienced decoder.

It becomes necessary, then, to devise a code which cannot be deciphered by such simple means. One such code is based on the binary number system and on a variation of the system of letter representation we discussed in the last section.

Using the telegraph code, we can represent any message by a definite sequence of five-digit combinations of zeros and ones. Suppose we set up in advance some absolutely arbitrary sequence of such five-digit binary numbers. Such a sequence, intended for coding a text, is called a *scale*. We make two copies of the scale, writing it as a combination of holes in a special paper tape (fig. 3), in which every row across on the tape contains some five-digit combination, a punched hole representing a one, and the absence of a hole representing a zero.

---

4. In addition to the coding system we have constructed, there is a widely accepted coding system called the Morse code, which also relies on representations of letters using combinations of two symbols—in this case, dots and dashes. We shall not discuss the details of the Morse code system here.

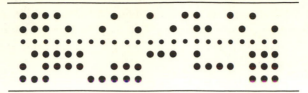

Fig. 3

We keep one copy of the scale and send the other to the person with whom we have a telegraphic connection. We now combine our message with the "arbitrarily" prepared scale in the following way: We "combine" the first five-digit number (the first letter) of the message with the first number of the scale, the second number of the message with the second number of the scale, and so on, "adding" in columns under the rules

$$0 \oplus 0 = 0, \quad 1 \oplus 0 = 0 + 1 = 1, \quad 1 \oplus 1 = 0 \, ;$$

that is, without carrying the sum of two units into the next place. The operation $\oplus$ is called "addition modulo 2"; clearly, such a method of combining two binary numbers yields a 0 digit in each place in which the corresponding digits of the two numbers are equal and a 1 in each place in which they are not. The result of such a combination of the text and the arbitrary scale can then be transferred as a sequence of electrical signals to our addressee. To restore the original message he need only add the same scale to the text in the manner described above.

The whole process can be described as follows:

1. text $\oplus$ scale = coded text;
2. coded text $\oplus$ scale = text $\oplus$ scale $\oplus$ scale = text.

It is not hard to see that for the purpose of sending a single message, this code is no better than the letter representation code of the last section; the scale serves only to permute the numbers which are assigned to our twenty-seven symbols. But when this code is used to send many different messages (using many different scales), the would-be decoder is faced with the task of breaking a new code with each message, even though no added hardship is imposed on those who know the code and its scaling principle. The code is far from "perfect," however, since an adversary with unlimited resources, even if he never discovered the scaling principle, could, in theory, break each new code in the same way as the letter representation code can be broken.

This entire process can easily be made automatic with the attachment of an apparatus that would perform the operation of combining message and scale to the transmitter, along with a similar apparatus to the receiver. The telegraph operators serving the line need not even know such mechanisms are present.

Of course, the binary system is especially convenient here because each number, when "added" to itself, yields a "sum" of zero, making the coding and decoding operations identical.

12. A Few Words about Computers

We have been speaking of the use of the binary system in a comparatively old province of technology, telegraphy (the first telegraphic apparatuses based on transmission of electrical signals via conductors appeared in the 1830s). We shall now consider one of the newest applications of the binary system—computers. But first we must discuss, although in very general terms, just what an electronic computer is.

The history of the development of computer technology is very lengthy and at the same time very short. The first devices designed to simplify the work of computation appeared long ago. For example, ordinary calculators were used for accounting purposes over four thousand years ago. Still, genuine "machine mathematics" arose no more than twenty-five years ago, when the first high-speed computers based on modern electronic technology (radio tubes and later transistors) appeared. In the short time the technology of computers achieved striking success. Modern computers work at speeds up to millions of operations per second; in other words, they perform in one second as many operations as an experienced human armed with a desk calculator can perform in several months. These machines have allowed us to solve problems which are so complex that solutions by hand would have been out of the question. For example, a modern computer is capable of solving a system of several hundred simultaneous linear equations with the same number of unknowns. A human "computer" armed with a pencil, paper, and desk calculator could not cope with such a problem in a lifetime.

When computers are mentioned in popular literature, we find expressions such as "the machine that solves complex equations," "the machine that plays chess," or "the machine that translates from one language to another." This can give the false impression that each such function—solving equations, playing chess, translating, and the like—is done by a specific machine built only for that purpose. However, all these problems and more—both mathematical (solving equations, constructing tables of logarithms, and so forth) and nonmathematical (translating a text or playing chess)—can be solved by one machine, the so-called universal computing machine. Strictly speaking, every such machine can perform only a very limited number of elementary operations: adding and multiplying numbers and storing the results in the machine's "memory," comparing numbers, choosing the largest or the smallest of two or more numbers, and the like. However, the solutions of the most diverse and complicated problems can be reduced to sequences (possibly very long) of such elementary operations. Such a sequence of operations is defined by a "program." Thus, variety in the problems which can be solved by a universal computing machine leads to variety in the programs fed into this machine.

In doing computations by hand or by computer, we must agree on some number system. When working with pencil and paper, we, of course, use the decimal system to which we are accustomed. However, the decimal system is hardly suitable for electronic computers. Such machines have a decided preference for the binary system. We shall now attempt to find the reasons for this.

13. Why Electronic Machines "Prefer" the Binary System

When we perform a computation by hand, we write the numbers on paper in pencil or pen. For a machine, however, some other method of storing the numbers with which it is operating is needed.

To clarify this problem, consider, for the moment, not a computing machine, but a far simpler apparatus—an ordinary counting device (electric meter, gas meter, taxi meter, and so on). Every such counter is composed of several discs, each of which can be situated in one of ten positions, corresponding to the digits from 0 to 9. It is clear, then, that an apparatus consisting of k such discs can serve to store one of 10^k different numbers, from 0 to $10^k - 1$. Such a counter could very well be used for computation; that is, it could be used not only to store numbers but also to perform arithmetical operations.

In general, a counter suited to a number system with base p is a system of discs, each of which has p different positions. In particular, the apparatus with which binary numbers could be stored should contain a number of objects, each of which would have two possible positions. It is clear that we need not use discs as the counting apparatus. In principle, a counter can consist of any collection of convenient elements, the only requirement being that each of the elements be able to take on as many stable conditions as there are units in the base of the number system being used.

A counter employing a system of wheels or any other mechanical apparatus changes its state relatively slowly. The speed with which modern computers work—millions of operations per second—is possible only because these machines work electronically rather than mechanically. Such machines are practically devoid of inertia and, therefore, can change their state within a time interval of a millionth of a second.

Electronic elements (vacuum tubes, transistors) typically have two stable conditions. For example, an electric bulb can be "on" (when current is passing through it) or "off" (when current is not passing through it). Semiconductors, now widely used in computer technology, operate by the same principle. This property of electronic elements is the basic reason why the binary system has proved to be the most convenient one for computers.

The input data for solving a problem is usually given in the conventional decimal system. Therefore, so that a machine based on the binary system can use the data, we must translate it into binary representation, a language that the machine's arithmetical apparatus can "understand." Such a translation is simple to accomplish automatically, of course. We also want the results of the computer's computations to be written in the decimal notation. Therefore, the computer generally must translate the result from the binary system into the decimal system.

Computers sometimes use a combined binary-decimal system. In this system, a number is first written in the ordinary decimal system, and then each of its digits is represented, using zeros and ones, in the binary system. In this manner, the binary-decimal system represents every number as several groups of zeros and ones. For example, the number

$$2593$$

is written in the binary-decimal system as

$$0010\ 0101\ 1001\ 0011\ .$$

In comparison, the binary representation of the same number is

$$101000100001\ .$$

Let us see how a computer based on the binary number system performs arithmetic operations. The basic operation which we should consider is addition, since multiplication reduces to iterated addition, subtraction reduces to addition of negative numbers, and, finally, division reduces to iterated subtraction. In turn, the addition of multi-digit numbers reduces to performing the appropriate addition place by place.

The addition of two binary numbers place by place can be described as follows.[5] Let a be the digit in a given place in the first summand; b, the digit in the same place in the second summand; and c, the digit which we have to carry from the preceding place (where, we are assuming, the addition has already taken place). To perform the addition in the given place is to indicate which digit (0 or 1) needs to be written as the "sum" and which digit must be carried to the next place.

We denote the digit that must be written in the given place of the sum by the letter s, and the value that we have to carry to the next place by the letter t. Since each of the quantities a, b, c, s, and t can take only a value of 0 or 1, all the possible variants involved are contained in the following table:

a	1	1	1	1	0	0	0	0
b	1	1	0	0	1	1	0	0
c	1	0	1	0	1	0	1	0
s	1	0	0	1	0	1	1	0
t	1	1	1	0	1	0	0	0

Thus, so that a computer can add two numbers written in the binary system, for every place there must exist an apparatus having three inputs corresponding to the values a, b, and c, and two outputs corresponding to the values s and t. Let us assume, as it usually occurs in electronic machines, that 1 is represented by the presence of current in the given input or output and 0 by its absence. The apparatus under consideration, called a *single-digit adder*, should work in an analogous way with the table above, that is, so that if there is no current in any of the three inputs, there will be none in either of the outputs; if there is current in a, but not in b or c, there should be current in s and none in t, and so on. An apparatus working by this scheme is not hard to construct using vacuum tubes or transistors.

14. One Remarkable Property of the Ternary System

In any evaluation of the "convenience" of a given number system, at least two criteria come into play: the simplicity of arithmetic computation in the system, and what is referred to as the "economy" of the

5. We speak here of the ordinary arithmetic addition and not the addition modulo 2 mentioned in section 11, in connection with a coded text. However, addition modulo 2 also has an essential place in the operation of a computer.

system. Economy is measured by the quantity of numbers that can be expressed in the number system with some arbitrary number of symbols.

Let us clarify this with an example. In order to write 1,000 numbers (from 0 to 999) in the decimal system, we need 30 "symbols" (10 digits for each place). In the binary system, we can write 2^{15} different numbers using 30 "symbols" (since, for every binary place, we need only 2 digits, 0 and 1, and so, with 30 symbols, we can write numbers containing up to 15 binary places). But

$$2^{15} > 1000 \; ;$$

therefore, using 15 binary places, we can write more different numbers than we can with three decimal places. In this sense, the binary system is more economical than the decimal system.

But which of all the number systems is the most economical? Let us consider the following concrete problem. Suppose we have at our disposal 60 symbols. We can separate them into 30 groups of 2 elements each, writing any number in the binary system using no more than 30 binary places, that is, 2^{30} numbers. We can also divide these 60 symbols into 20 groups of 3 elements each and, using the ternary system, write 3^{20} different numbers. Furthermore, by separating the 60 symbols into 15 groups of 4 elements each, we can apply the base 4 system and write 4^{15} numbers, and so forth. In particular, if we used the decimal system (that is, separating all the symbols into 6 groups of 10 elements each), we could write 10^6 numbers, but if we used the sexagesimal (base sixty) system, 60 symbols would allow us to write only 60 numbers. Let us find out which of the possible systems is the most economical; that is, which one allows us to write the greatest quantity of numbers using only 60 symbols. In other words, we are asking which of the numbers

$$2^{30}, \, 3^{20}, \, 4^{15}, \, 5^{12}, \, 6^{10}, \, 10^6, \, 12^5, \, 15^4, \, 20^3, \, 30^2, \, 60$$

is the largest. It can be verified by calculation that the largest number is 3^{20}. We first show that

$$2^{30} < 3^{20} \; .$$

Since $2^{30} = (2^3)^{10} = 8^{10}$ and $3^{20} = (3^2)^{10} = 9^{10}$, we can write our inequality in the form

$$8^{10} < 9^{10} \; .$$

In this form, our result is obvious.

Furthermore,

$$4^{15} = (2^2)^{15} = 2^{30} .$$

Thus, by what we have just shown,

$$3^{20} > 4^{15} .$$

It is easy to verify that the following chain of inequalities is valid:

$$4^{15} > 5^{12} > 6^{10} > 10^6 > 12^5 > 15^4 > 20^3 > 30^2 > 60 .$$

Thus, the ternary system has turned out to be most economical, with the binary and base four systems next best.

This result is in no way dependent on the fact that we were considering 60 symbols. We chose this example only because a group of 60 symbols is easily divided into groups of 2, 3, 4, and so forth.

In the general case, if we employ n symbols and use some number x for the base of the number system, then we can use n/x places, and the quantity of numbers that we can write will be equal to

$$x^{n/x} .$$

Consider this expression as a function of the variable x, taking not only integral but any (fractional, irrational) positive values. It is possible to find the value of x at which the function achieves its maximum. The function has a maximum at e, an irrational number which is the base of the so-called system of natural logarithms and which plays an important role in the most diverse questions of higher mathematics.[6] The number e is approximately equal to

$$2.718281828459045 \ldots .$$

6. For the reader familiar with the elements of differential calculus, we give the corresponding calculation. A necessary condition for a function $y(x)$ to achieve its maximum at a point x_0 is that the derivative of the function be zero at that point. In the given case,

$$y(x) = x^{n/x} .$$

The derivative is equal to

$$\frac{dy}{dx} = \frac{d}{dx} [(x)^{n/x}] = \frac{d}{dx} (e^{n \ln x/x})$$

$$= \left(\frac{n}{x} \cdot \frac{1}{x} - \frac{n \ln x}{x^2} \right) e^{n \ln x/x}$$

$$= \frac{n}{x^2} (1 - \ln x) e^{n \ln x/x}$$

$$= \frac{n}{x^2} (1 - \ln x) x^{n/x} .$$

The closest integer to e is 3, which serves as the base for the most economical number system.

The graph of the function $y = (x)^{n/x}$ is given in figure 4. (Note, however, that the x- and y-axes have different scales.)

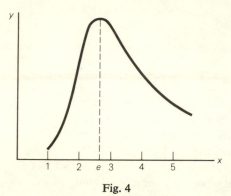

Fig. 4

The economy of a number system is a significant property from the standpoint of its use in computer technology. For this reason, although the use of the ternary system in place of the binary system in computers involves some difficulties in construction (one must use elements that can exist in three rather than in two stable conditions), the ternary system has already been tried in several existing computers.

15. On Infinite Number Representations

Up to this point, we have considered number system representations only of the integers. It is natural, however, to pass from the decimal notation of whole numbers to decimal representation of fractions. To do so, we must consider not only the nonnegative powers of 10 (1, 10, 100, and so on), but negative powers (10^{-1}, 10^{-2}, and so on), and compose combinations in which we use these negative powers as well as the others. For example, the expression 23.581 stands for

$$2 \cdot 10^1 + 3 \cdot 10^0 + 5 \cdot 10^{-1} + 8 \cdot 10^{-2} + 1 \cdot 10^{-3}.$$

Fractions are conveniently represented as points on a line. We take a

---

Setting the derivative equal to zero, we obtain

$$\ln x = 1, \text{ that is, } x = e.$$

Since the derivative dy/dx is positive to the left of $x = e$ and negative to the right, we can use a well-known theorem of differential calculus to show that our function has a maximum at that point.

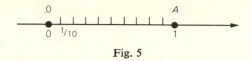

Fig. 5

line and choose a fixed point O (the origin of the line), a positive direction (to the right), and a unit of measure, the line segment OA (fig. 5). We take the point O to stand for the number zero, and the point A to stand for the number 1. Having laid the segment OA to the right of the point O two, three, etc. times, we obtain points which represent the numbers 2, 3, and so on. In this way we can represent all the integers on a line. To represent fractions containing tenths, hundredths, and so on, we need only divide the segment OA into ten, one hundred, and so forth, equal parts and use these smaller units of length. We can thus measure off points on the line corresponding to all possible numbers of the form

$$a_k a_{k-1} \cdots a_1 a_0 . b_1 b_2 \cdots b_n \;;$$

that is, all possible finite decimal representations. In doing so, of course, we do not obtain all the points of the line. For example, the endpoint of a segment of the same length as a diagonal of the unit square (the square with side 1) does not correspond to any finite decimal representation, since the ratio of the length of a square's diagonal to the length of its side is irrational.

If we want *each* point of the line to correspond to some number, we shall have to use not only finite, but infinite decimal representations. Let us clarify the meaning of this last statement.

In order to make every point of the line correspond to some (infinite) decimal representation, we proceed in the following manner. For convenience, we shall speak only about a part of the whole line, the line segment OA—our unit interval. Let x be some point on this line segment. We divide OA into 10 equal parts and number the parts using the digits from 0 to 9. We denote the number of the section in which x lies by b_1. We now divide this smaller segment into 10 parts, numbering these parts in the same way, and denoting the number (0 to 9) of the smaller section by b_2. We subdivide further in the same way, continuing the process indefinitely. As a result, we obtain a sequence of digits $b_1, b_2, \ldots, b_n, \ldots$, which we write in the form

$$.b_1 b_2 \cdots b_n \cdots ,$$

and which we call the *infinite decimal representation* (or *infinite decimal expansion*) corresponding to the point x. If we break off this expansion at some point, we get an ordinary (finite) decimal representation $.b_1b_2 \cdots b_n$, which defines the position of the point x only approximately (with an accuracy of a (10^n)th part of the unit interval).

In this way, we have assigned to each number x between 0 and 1 an infinite decimal expansion. The correspondence can be extended to the entire line, for if the number y lies between the integers n and $n + 1$, the number $x = y - n$ lies between 0 and 1 and thus has some decimal representation

$$x = .b_1b_2 \cdots b_n \cdots .$$

If the integer n has the decimal representation

$$a_k a_{k-1} \cdots a_1 a_0 ,$$

then y can be written

$$y = n + x = a_k a_{k-1} \cdots a_1 a_0 . b_1 b_2 \cdots b_n \cdots .$$

It is not hard to see that some uncertainty inevitably arises from this. In particular, having divided the segment OA into 10 parts, we must consider, for example, the point on the boundary between the first and the second parts. We can consider it to be both in the first section (having number 0) and in the second (having number 1). In the first case, continuing the process of successive divisions, we will discover that the chosen point is in the rightmost (having number 9) of all the parts into which we divide the segment at each step, that is, we obtain the infinite fraction

$$0.0999\ldots,$$

while in the second case the point will be in each of the sections which have number 0, that is, yielding the fraction

$$0.1000\ldots.$$

Here we have obtained two infinite representations corresponding to one and the same point. The same thing will occur at any other boundary point (between two segments) in any of the successive divisions. For example, the fractions

$$0.125000\ldots \quad \text{and} \quad 0.124999\ldots$$

represent one and the same point.

We can avoid this ambiguity by agreeing to think consistently of every boundary point as belonging either to the rightmost or the leftmost of the partial segments which it bounds. In other words, we can eliminate either all fractions consisting of "infinitely repeating" zeros, or all fractions consisting of "infinitely repeating" nines.

If we introduce such a restriction, we can represent each point of the line by a unique infinite decimal expansion.

That we have successively divided the partial segments into 10 parts is, of course, immaterial. Instead of 10, we could have used some other number, say 2, dividing each partial segment in half. In this way we can represent each point of the line by an infinite sequence $b_1, b_2, \ldots, b_n, \ldots$ of zeros and ones, which we write in the form

$$(0.b_1 b_2 \cdots b_n \cdots)_2$$

and call an *infinite binary representation* (or *expansion*). If we cut off this sequence at some place, we get the finite binary representation

$$(0.b_1 b_2 \cdots b_n)_2 \; ;$$

that is, the number

$$b_1 \cdot 1/2 + b_2 \cdot 1/2^2 + \cdots + b_n \cdot 1/2^n \, ,$$

approximating the point under consideration to within a (2^n)th part of the unit interval.

Infinite decimal expansions, with which we can represent all the points of the line, are a convenient tool in the construction of the theory of real numbers, which is fundamental in many aspects of higher mathematics. However, any other type of infinite expansions (binary, ternary, and so on) can be used with equal success.

Before concluding, let us consider the following instructive problem. We take a line segment OA, divide it into three equal parts, and reject its middle part (we consider the points of division themselves to be members of the middle part—that is, they are also rejected; fig. 6). We

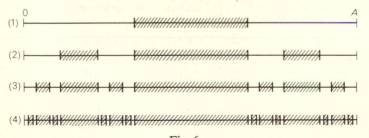

Fig. 6

further divide each of the two remaining parts into three equal parts and reject the center segments. After this there remain four small pieces, from each of which we again take the middle third. We continue this process indefinitely. How many points of the segment OA will remain undeleted?

At first glance we might say that only the endpoints O and A will remain. This conclusion is supported, it would seem, by the following reasoning. We compute the sum of the lengths of all the segments deleted by the above process. (We recall that we took the length of the entire segment OA to be equal to 1.) At the first step we rejected a segment of length 1/3, at the second step two segments of length 1/9, at the third four segments of length 1/27, and so on. The sum of the lengths of all the deleted segments is equal to

$$1/3 + 2/9 + 4/27 + \dots.$$

This is an infinite geometric progression with first term 1/3 and ratio 2/3. By the well-known formula, its sum is equal to

$$\frac{1/3}{1 - 2/3} = 1.$$

Thus, the sum of the lengths of the deleted segments is exactly equal to the length of the original segment OA!

And yet the above process leaves—besides O and A—an infinite number of undeleted points. To see this, we represent each point of the unit segment OA by an infinite ternary expansion. Each such representation consists of zeros, ones, and twos. We claim that the process of deleting the "middle third" leaves behind exactly those points which correspond to ternary expansions containing no ones (composed entirely of zeros and twos). In the first step we deleted the middle third of the unit interval, that is, those points which correspond to ternary expansions having a one in the first place. In the second step we deleted the middle third again, removing the expansions which have a one in the second place, and so forth. (Here we delete those points that can be represented by two ternary expansions if one of these expansions contains a one. For example, the endpoint of the first third of the line segment OA, the number 1/3, can be represented by the ternary expansions

$$0.1000\dots$$

and

$$0.0222\ldots\ ;$$

this point we delete.) And so, the process leaves exactly those points which correspond to ternary expansions consisting only of zeros and twos. But there are infinitely many such numbers! Consequently, besides the endpoints, there will still remain infinitely many undeleted points. For example, the point that corresponds to the representation

$$0.020202\ldots$$

(the ternary expansion of the number 1/4) will remain. The infinite ternary representation 0.020202... actually signifies the sum of the geometric progression

$$2\cdot3^{-2} + 2\cdot3^{-4} + 2\cdot3^{-6} + \ldots,$$

which, by the formula, is equal to

$$\frac{2/9}{1-1/9} = \frac{2/9}{8/9} = 1/4\,.$$

By using the following geometric argument, we can persuade ourselves that the point 1/4 will not be deleted. The point 1/4 divides the whole interval [0, 1] in a ratio of 1:3. After the removal of the segment [1/3, 2/3], the point 1/4 remains in the half-open interval [0, 1/3), which it divides in a ratio of 3:1. After the second deletion it remains in the open interval (2/9, 1/3), which it divides in a ratio of 3:1, and so on. At no step will the point 1/4 be removed.

Thus, it turns out that the process of deleting the "middle third" leads to a set of points which, although it "takes up no space at all" on the line segment (since the sum of the lengths of the deleted segments is equal, as we have made clear, to one), contains infinitely many points.

This set of points possesses other interesting properties; however, to study them would require an exposition of concepts beyond the scope of our little book. Thus, we end here.